Volunteer Amateur Radio Operator Hospital Orientation

October, 2015
Updated July, 2019

Duane Mariotti, WB9RER

EMERGENCY RADIO OPERATOR HOSPITAL ORIENTATION

INTRODUCTION

Amateur radio has and can support hospital emergency communications during disasters. However, a radio operator supporting a hospital requires knowledge of the hospital environment, regulations and procedures to best serve the hospital customer.

All new hospital staff including physicians, contractors, volunteers and employees typically attend hospital orientation. Hospital orientation focuses on operations, regulatory requirements and safety. This book is a self-paced orientation geared specifically for the hospital emergency amateur radio operator. It is one component of a complete orientation for the hospital emergency amateur radio operator.

This book also provides an emergency amateur radio operator with "big picture" situational awareness to consider prior to supporting a hospital during an emergency.

EMERGENCY RADIO OPERATOR HOSPITAL ORIENTATION

Table of Contents

- Hospital Orientation Introduction
- Hospital Number and Size (Beds)
- Hospital Accreditation
- Trauma Centers – Adult and Pediatric
- Neonatal Intensive Care Units (NICU)
- Emergency Departments – Burn Centers
- Hospital Orientation Overview
- Volunteer Professionalism
- Regulatory Overview and HIPAA
- Hospital Safety – Hospital Emergency Codes – Fire Safety
- Safety – Emergency Power – Security – Disaster Preparedness
- Infection Prevention
- Kaiser Permanente
- Quiz

HOSPITAL ORIENTATION INTRODUCTION FOR AMATEUR RADIO OPERATOR

WE ARE SUPPORTING HOSPITAL EMERGENCY COMMUNICATIONS
- WE SHOULD KNOW SOMETHING ABOUT HOSPITAL OPERATIONS

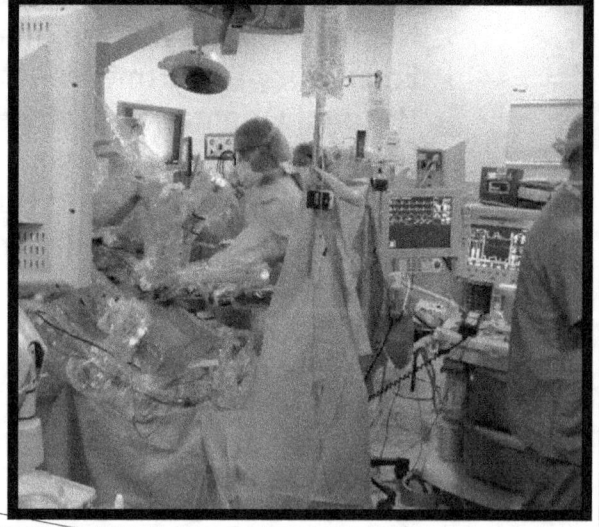

HOSPITAL ORIENTATION

All employees, volunteers and contractors performing service in a hospital must be orientated per various regulations.

These orientations may encompass hours to over four weeks depending on the role in the hospital.

Hospital orientation is typically configured primarily to assist with interactions and activities related to patient care.

HOSPITAL ORIENTATION

Kaiser Permanente Amateur Radio Network (KPARN) Radio Operators by definition are limited to public spaces and specific secure locations such as the Hospital Command Center (HCC). It is not the intent of the KPARN structure for radio operators to be in patient care areas or situations.

This orientation training is specifically tailored to the Amateur Radio Operator and their limited but critical role in support of hospital emergency management in non-clinical settings.

HOSPITAL ORIENTATION
HOSPITALS ARE SMALL CITIES
HOSPITAL DEPARTMENT NAMES

- Hospital Administration
- Nursing Administration
- Admitting
- Dietary
- Maintenance / Engineering
- Laundry
- Anesthesia
- Surgical Services – Operating Room
- Medical Surgical Unit – Nursing Unit
- Step Down Nursing Unit
- Telemetry Nursing Unit
- Intensive Care Unit
- Nuclear Medicine
- Medical Imaging
- Clinical Laboratory
- Blood Bank
- Pathology
- Materials Management - Purchasing
- Pharmacy – In Patient and Out Patient
- Nursery
- Patient Transportation
- Housekeeping - EVS
- Quality Improvement
- Risk Management
- Security
- Emergency Department
- Post Anesthesia Care Unit – Recovery
- Pain Management
- Central Service / Sterile Processing
- Interpreter
- Medical Records
- Infection Control
- Accounts Payable
- Patient Accounting
- Labor and Delivery – Birth Center
- Occupational Therapy
- Cardiology - Stress/EKG Department
- Social Services
- Community Relations
- Respiratory Therapy
- Physical Therapy
- Hemodialysis
- Pediatrics

HOSPITAL ORIENTATION

- APPROX. 17,000 HOSPITALS WORLD WIDE
- APPROX. 6,300 HOSPITALS (PSYCH, ACUTE, REHAB, FEDERAL) IN USA
- APPROX. 5,300 ACUTE CARE HOSPITALS IN USA
- 2017 – 800,000 ACUTE CARE BEDS IN USA **(1975 THERE WERE 1.5M BEDS)**
- 362 ACUTE CARE HOSPITALS IN CALIFORNIA, SEVEN PERCENT OF USA HOSPITALS
- 73,000 ACUTE CARE BEDS IN CALIFORNIA, NINE PERCENT OF USA HOSPITAL BEDS
- **CALIFORNIA IS TWELVE PERCENT OF USA POPULATION**

2017 DATA - KAISER FAMILY FOUNDATION

NUMBER OF ACUTE CARE HOSPITALS BY STATE

STATE	# HOSP	STATE	# HOSP	STATE	# HOSP	STATE	# HOSP
TEXAS	365	ALABAMA	92	MINNESOTA	55	MAINE	19
CALIFORNIA	340	VIRGINIA	90	COLORADO	53	IDAHO	17
FLORIDA	215	OKLAHOMA	90	MARYLAND	52	MONTANA	16
NEW YORK	190	MISSOURI	84	ARKANSAS	52	HAWAII	14
PENNSYLVANIA	178	WISCONSIN	76	IOWA	39	NEW HAMPSHIRE	14
OHIO	145	KENTUCKY	75	OREGON	37	WYOMING	14
ILLINOIS	143	NEW JERSEY	74	WEST VIRGINIA	37	RHODE ISLAND	11
GEORGIA	113	MASSACHUSETTS	73	NEW MEXICO	37	NORTH DAKOTA	10
NO CAROLINA	109	MISSISSIPPI	71	UTAH	35	ALASKA	10
MICHIGAN	105	ARIZONA	70	CONNECTICUT	34	DELAWARE	8
LOUISIANA	105	WASHINGTON	64	NEVADA	30	DC	7
TENNESSEE	103	SO CAROLINA	63	NEBRASKA	27	VERMONT	7
INDIANA	97	KANSAS	57	SOUTH DAKOTA	24		

THERE ARE 81 HOSPITALS (9-1-1) IN LA COUNTY. LA COUNTY HAS MORE HOSPITALS THAN 34 STATES. THERE ARE 155 HOSPITALS IN THE FIVE COUNTIES OF SOUTHERN CALIFORNIA. THIS IS MORE HOSPITALS THAN 46 STATES.

AMERICAN HOSPITAL DIRECTORY 2017

ACUTE CARE HOSPITAL BEDS PER STATE

STATE	# HOSP	# BEDS	STATE	# HOSP	# BEDS	STATE	# HOSP	# BEDS	STATE	# HOSP	# BEDS
CALIFORNIA	340	73,896	MISSOURI	84	16,897	MISSISSIPPI	71	10,154	SOUTH DAKOTA	24	2,756
FLORIDA	215	60,153	INDIANA	97	16,087	CONNECTICUT	34	8,854	HAWAII	14	2,548
TEXAS	365	58,878	ALABAMA	92	15,716	COLORADO	53	8,370	IDAHO	17	2,456
NEW YORK	190	56,469	MASSACHUSETTS	73	15,142	ARKANSAS	52	8,025	RHODE ISLAND	11	2,424
PENNSYLVANIA	178	35,843	LOUISIANA	105	14,511	KANSAS	57	6,422	NEW HAMPSHIRE	14	2,237
ILLINOIS	143	30,167	ARIZONA	70	13,867	IOWA	39	6,319	MONTANA	16	2,193
OHIO	145	27,782	KENTUCKY	75	13,863	OREGON	37	6,214	DC	7	2,151
MICHIGAN	105	23,702	SO CAROLINA	63	11,665	WEST VIRGINIA	37	5,964	DELAWARE	8	2,042
GEORGIA	113	22,243	WISCONSIN	76	11,215	NEVADA	30	5,640	NORTH DAKOTA	10	2,006
NO CAROLINA	109	22,144	MARYLAND	52	10,933	UTAH	35	4,588	WYOMING	14	1,240
NEW JERSEY	74	20,229	MINNESOTA	55	10,664	NEBRASKA	27	4,199	ALASKA	10	1,231
TENNESSEE	103	18,674	OKLAHOMA	90	10,524	NEW MEXICO	37	4,006	VERMONT	7	828
VIRGINIA	90	17,849	WASHINGTON	64	10,336	MAINE	19	2,917			

TEXAS HAS MORE HOSPITALS THAN CALIFORNIA. CALIFORNIA HAS THE MOST AVAILABLE ACUTE CARE BEDS IN USA.

AMERICAN HOSPITAL DIRECTORY 2017

Amateur Radio Hospital Orientation

LARGEST ACUTE CARE HOSPITALS – USA
(BY NUMBER OF BEDS)

1. Florida Hospital Orlando (Fla.) — 2,473 BEDS.
2. New York-Presbyterian Hospital (New York City) — 2,428 BEDS.
3. Jackson Memorial Hospital (Miami) — 1,750 BEDS.
4. Methodist Hospital (San Antonio) — 1,570 BEDS.
5. Baptist Medical Center (San Antonio) - 1,563 BEDS

31. Cedars-Sinai Medical Center (Los Angeles) — 947 BEDS.

In USA 27 Hospitals have 1,000 or more beds
www.beckershospitalreview.com – 2017 DATA

MGM GRAND IN LAS VEGAS HAS 6,852 ROOMS

LARGEST ACUTE CARE HOSPITALS–CALIFORNIA

NAME	CITY	COUNTY	BEDS
CEDARS SINAI MC	LOS ANGELES	LOS ANGELES	886
SANTA CLARA VALLEY MC	SAN JOSE	SANTA CLARA	731
COMMUNITY REGIONAL MC	FRESNO	FRESNO	685
LAC+USC MEDICAL CENTER	LOS ANGELES	LOS ANGELES	676
SHARP MEMORIAL HOSPITAL	SAN DIEGO	SAN DIEGO	656
UC DAVIS MEDICAL CENTER	SACRAMENTO	SACRAMENTO	625
HUNTINGTON MEMORIAL HOSPITAL	PASADENA	LOS ANGELES	619
STANFORD HEALTH CARE	PALO ALTO	SANTA CLARA	613
TORRANCE MEMORIAL MC	TORRANCE	LOS ANGELES	610
UCSF MEDICAL CENTER	SAN FRANCISCO	SAN FRANCISCO	590
JOHN MUIR MC-WALNUT	WALNUT CREEK	CONTRA COSTA	554
RADY CHILDREN'S HOSPITAL	SAN DIEGO	SAN DIEGO	524
GROSSMONT HOSPITAL	LA MESA	SAN DIEGO	524
PIH HEALTH HOSPITAL	WHITTIER	LOS ANGELES	523
SUTTER MC	SACRAMENTO	SACRAMENTO	523
ADVENTIST HEALTH	GLENDALE	LOS ANGELES	515
CHILDREN'S HOSPITAL OF LA	LOS ANGELES	LOS ANGELES	495
SCRIPPS MERCY HOSPITAL	SAN DIEGO	SAN DIEGO	482
SANTA BARBARA COTTAGE HOSPITAL	SANTA BARBARA	SANTA BARBARA	481
RIVERSIDE COMMUNITY HOSPITAL	RIVERSIDE	RIVERSIDE	478

OSHPD 2019

HOSPITAL ACCREDITATION

Accreditation, per Mosby Medical Dictionary, is a process whereby recognition is granted to a healthcare institution for demonstrated ability to meet established standards.

Hospitals may have numerous accreditations. The primary accreditation is The Joint Commission (TJC) or equivalent which inspects hospital every three-years for entire safe patient experience. Failure affects insurance reimbursement.

Other accreditations include laboratory (CAP), mammography, ultrasound, physical therapy, rehab medicine, cardiac (STEMI), stroke and many, many others.

Knowledge of local accreditations of most value to a hospital emergency radio operator post disaster is emergency department volume, **burn care, trauma care, pediatric care** and possibly care of the newborn (neonatal).

USA TRAUMA CENTERS

TOTAL OF 397 ADULT AND 140 PEDIATRIC LEVEL ONE TRAUMA CENTERS IN USA.
- 225 STATE VERIFIED LEVEL ONE ADULT TRAUMA CENTERS
- 172 ACS VERIFIED LEVEL ONE ADULT TRAUMA CENTERS
- 82 STATE VERIFIED LEVEL ONE PEDIATRIC TRAUMA CENTERS
- 58 ACS VERIFIED LEVEL ONE PEDIATRIC TRAUMA CENTERS

TOTAL OF 536 ADULT AND 110 PEDIATRIC LEVEL TWO TRAUMA CENTERS IN USA.
- 345 STATE VERIFIED LEVEL TWO ADULT TRAUMA CENTERS
- 191 ACS VERIFIED LEVEL TWO ADULT TRAUMA CENTERS
- 61 STATE VERIFIED LEVEL TWO PEDIATRIC TRAUMA CENTERS
- 49 ACS VERIFIED LEVEL TWO PEDIATRIC TRAUMA CENTERS

ACS = AMERICAN COLLEGE OF SURGEONS
AMERICAN TRAUMA SOCIETY 2019

LEVEL ONE TRAUMA CENTER

Level One Trauma Center is defined by the American College of Surgeons (ACS) as a comprehensive regional resource that is a tertiary care facility central to the trauma system. A Level One Trauma Center is capable of providing **total care for every aspect of injury – from prevention through rehabilitation.**

Level One Trauma Center has 24-hour in-house coverage by general surgeons, and **prompt availability** of care in specialties such as orthopedic surgery, neurosurgery, anesthesiology, emergency medicine, radiology, internal medicine, plastic surgery, oral and maxillofacial, pediatric and critical care.

ADULT LEVEL 1 TRAUMA CNTR SOUTHERN CALIFORNIA

- **UC IRVINE MC** – ORANGE COUNTY
- **SCRIPPS MERCY MC** – SAN DIEGO COUNTY
- **UC SAN DIEGO MC** – SAN DIEGO COUNTY
- **UCLA MC** – LOS ANGELES COUNTY
- **CEDARS-SINAI MC** – LOS ANGELES COUNTY
- **HARBOR UCLA MC** – LOS ANGELES COUNTY
- **LAC + USC MC** – LOS ANGELES COUNTY
- **LOMA LINDA UNIVERSITY MC** – SAN BERNADINO COUNTY

American College of Surgeons 2019

LEVEL TWO TRAUMA CENTER

A Level Two Trauma Center is able to **initiate definitive care** for all injured patients.

Level Two Trauma Center 24-hour **immediate coverage by general surgeons**, as well as coverage by the specialties of orthopedic surgery, neurosurgery, anesthesiology, emergency medicine, radiology and critical care.

Tertiary care needs such as cardiac surgery, hemodialysis and microvascular surgery may be referred to a Level One Trauma Center.

AMERICAN COLLEGE OF SURGEONS

ADULT LEVEL 2 TRAUMA CNTR SOUTHERN CALIFORNIA

Los Angeles	Long Beach Memorial MC
Los Angeles	Antelope Valley Hospital
Los Angeles	Dignity Health California Hospital MC
Los Angeles	Dignity Health St. Mary MC
Los Angeles	Dignity Health Northridge Hospital MC
Los Angeles	Henry Mayo Newhall Memorial Hospital
Los Angeles	Huntington Memorial Hospital
Los Angeles	Pomona Valley Hospital MC
Los Angeles	Providence Holy Cross MC
Los Angeles	St. Francis MC
Orange	Mission Hospital Regional MC
Orange	Orange County Global MC
Riverside	Desert Regional MC
Riverside	Riverside Community Hospital
Riverside	Riverside University Health System
San Bernardino	Arrowhead Regional MC
San Diego	Palomar MC
San Diego	Scripps Memorial Hospital
San Diego	Sharp Memorial Hospital

AMERICAN COLLEGE OF SURGEONS 2019

PEDIATRIC TRAUMA CENTERS SOUTHERN CALIFORNIA

LEVEL ONE
- LOMA LINDA UNIVERSITY MC
- CHILDREN'S HOSPITAL OF LOS ANGELES (CHLA)
- RADY CHILDREN'S HOSPITAL – SAN DIEGO

LEVEL TWO
- CEDARS–SINAI MC
- HARBOR–UCLA MC
- DIGNITY HEALTH NORTHRIDGE HOSPITAL
- LAC + USC MEDICAL CENTER
- UC IRVINE MC
- CHILDEN'S HOSPITAL ORANGE COUNTY (CHOC)
- LONG BEACH MEMORIAL MC – MILLER CHILDREN'S
- MISSION HOSPITAL REGIONAL MC – ORANGE COUNTY

AMERICAN COLLEGE OF SURGEONS 2019

NEONATAL INTENSIVE CARE

Level I neonatal care (**basic**), Well-newborn nursery: has the capabilities to provide neonatal resuscitation at every delivery. Evaluate and provide postnatal care to healthy newborn infants.

Level II neonatal care (**specialty**), Special care nursery:
- **Level IIA**: has the capabilities to Resuscitate and stabilize ill infants before transfer
- **Level IIB**: has the capabilities of a level IIA nursery and the additional capability to provide mechanical ventilation for brief durations (24 hours)

Level III (subspecialty) NICU
- **Level IIIA**: has the capabilities to provide sustained life support via conventional mechanical ventilation. Perform minor surgical procedures such as placement of central venous catheter.
- **Level IIIB**: has the capabilities to provide advanced respiratory support, advanced imaging, full range of on-site pediatric medical subspecialists and on-site pediatric surgical specialists.
- **Level IIIC**: is located within an institution that has the capability to provide cardiopulmonary bypass.

AMERICAN ACADEMY OF PEDIATRICS

"REGIONAL" OR LEVEL 3 NICU SOUTHERN CALIFORNIA

LOMA LINDA – LOMA LINDA – LEVEL 1 PED TRAUMA
RADY'S CHILDREN – SAN DIEGO – LEVEL 1 PED TRAUMA
CHILDRENS HOSPITAL OF LOS ANGELES (CHLA) – LEVEL 1 PED TRAUMA
CEDAR-SINAI MC – LOS ANGELES – LEVEL 2 PED TRAUMA
LAC+USC MC – LOS ANGELES – LEVEL 2 PED TRAUMA
HARBOR-UCLA MC – TORRANCE – LEVEL 2 PED TRAUMA
MILLER CHILDRENS – LONG BEACH – LEVEL 2 PED TRAUMA
UC IRVINE – IRVINE – LEVEL 2 PED TRAUMA
CHILDRENS HOSPITAL OF ORANGE COUNTY (CHOC) – LEVEL 2 PED TRAUMA
UCLA MATTEL – LOS ANGELES
KP LAMC – LOS ANGELES
UC SAN DIEGO – SAN DIEGO

CA DHCS

"COMMUNITY" OR LEVEL 2 NICU SOUTHERN CALIFORNIA

- LOS ANGELES COUNTY – 25
- ORANGE COUNTY – 7
- SAN DIEGO COUNTY – 6
- RIVERSIDE COUNTY – 4
- SAN BERNARDINO COUNTY – 4
- KERN COUNTY – 2

CA DHCS

BUSIEST USA ED'S ANNUAL PATIENT VISITS

1. Lakeland Regional MC – Florida – 217,000
2. Parkland – Dallas – 171,000
3. NYC Health & Hospitals – Bronx – 164,000
4. St. Joseph MC – Paterson, NJ – 163,000
5. WellStar Kennestone Hospital – Marietta, GA – 145,000
6. Banner Desert MC – Mesa, AZ – 145,000
7. NYC Health & Hospitals – Brooklyn – 139,000
8. Erlanger MC – Chattanooga, TN – 139,000
9. NYC Health & Hospitals – Elmhurst, NY – 137,000
10. Grady – Atlanta – 134,000

ED = Emergency Department

Becker's 2017

BUSIEST CALIFORNIA ED'S ANNUAL PATIENT VISITS

KP HOSPITAL - FONTANA	127,074
KP HOSPITAL - SOUTH SACRAMENTO	117,577
KP HOSPITAL - OAKLAND/RICHMOND	116,124
ANTELOPE VALLEY HOSPITAL	116,062
ADVENTIST HEALTH HANFORD	115,620
LAC+USC MEDICAL CENTER	113,509
VALLEY CHILDREN'S HOSPITAL	110,987
KP HOSPITAL - SACRAMENTO	106,973
KP HOSPITAL - ROSEVILLE	106,496
RIVERSIDE COMMUNITY HOSPITAL	105,795
KP HOSPITAL - ORANGE COUNTY - ANAHEIM	104,502
HOAG MEMORIAL HOSPITAL PRESBYTERIAN	100,275
SCRIPPS MERCY HOSPITAL	99,666

CA OSHPD 2017

BUSIEST SO. CALIFORNIA ED'S ANNUAL PATIENT VISITS

KP HOSPITAL - FONTANA	127,074
ANTELOPE VALLEY HOSPITAL	116,062
LAC+USC MEDICAL CENTER	113,509
RIVERSIDE COMMUNITY HOSPITAL	105,795
KP HOSPITAL - ORANGE COUNTY - ANAHEIM	104,502
HOAG MEMORIAL HOSPITAL PRESBYTERIAN	100,275
SCRIPPS MERCY HOSPITAL	99,666
KP HOSPITAL - SAN DIEGO - ZION	99,440
KP HOSPITAL - DOWNEY	95,820
PROVIDENCE HOLY CROSS MEDICAL CENTER	92,822

CA OSHPD 2017

HOSPITAL ORIENTATION
County Residents per Hospital Emergency Department (2019 DATA)

- "911 HOSPITALS" –
- COUNTY CERTIFIED EMS RECEIVING HOSPITALS

  - KERN COUNTY – 10 Hospitals – 0.9M RESIDENTS (90K/ED)
  - LOS ANGELES COUNTY – 81 – 10M RESIDENTS (124K/ED)
  - ORANGE COUNTY – 25 – 3.2M RESIDENTS (128K/ED)
  - RIVERSIDE COUNTY – 15 – 2.5M RESIDENTS (167K/ED)
  - SAN BERNARDINO CNTY – 16 – 2.2M RESIDENTS (138K/ED)
  - SAN DIEGO COUNTY – 18 – 3.3M RESIDENTS (183K/ED)

COUNTY EMS AGENCIES

ABA BURN CENTERS & ACS ADULT LEVEL ONE TRAUMA CENTERS

THERE ARE A TOTAL OF 67 AMERICAN BURN ASSOCIATION (ABA) VERIFIED BURN CENTERS IN USA – 33 OF THE 67 ARE ALSO ACS VERIFIED LEVEL ONE ADULT TRAUMA CENTERS

MARICOPA MC	PHOENIX	AZ	A & P	ESKENAZI HEALTH	INDIANAPOLIS	IN	A
COMMUNITY REGION MC	FRESNO	CA	A & P	THE U OF KANSAS	KANSAS CITY	KS	A & P
SANTA CLARA VALLEY MC	SAN JOSE	CA	A & P	BRIGHAM AND WOMENS	BOSTON	MA	A
LAC + USC MC	LOS ANGELES	CA	A & P	DETROIT RECEIVING	DETROIT	MI	A
UC IRVINE	ORANGE	CA	A & P	U OF MICHIGAN	ANN ARBOR	MI	A & P
UC DAVIS MC	SACRAMENTO	CA	A	HENNEPIN CNTY MC	MINNEAPOLIS	MN	A & P
UC SAN DIEGO MC	SAN DIEGO	CA	A & P	REGIONS HOSPITAL	ST PAUL	MN	A & P
U OF COLORADO	AURORA	CO	A	WAKE FOREST BAPTIST MC	WINSTON-SALEM	NC	A & P
MEDSTAR WASHINGTON	WASHINGTON	DC	A	METROHEALTH MC	CLEVELAND	OH	A & P
UF SHANDS	GAINESVILLE	FL	A	WEXNER MC OSU	COLUMBUS	OH	A
TAMPA GENERAL	TAMPA	FL	A & P	RHODE ISLAND HOSPITAL	PROVIDENCE	RI	A & P
JACKSON MEMORIAL	MIAMI	FL	A & P	MEMORIAL HERMANN	HOUSTON	TX	A
KENDALL REGIONAL MC	MIAMI	FL	A	PARKLAND MEMORIAL	DALLAS	TX	A & P
ORLANDO REGIONAL MC	ORLANDO	FL	A	U OF TEXAS MEDICAL	GALVESTON	TX	A
GRADY MEMORIAL	ATLANTA	GA	A & P	U OF UTAH	SALT LAKE CITY	UT	A & P
LOYOLA U MC	MAYWOOD	IL	A & P	U OF WISCONSIN	MADISON	WI	A & P
U OF IOWA	IOWA CITY	IA	A & P				

AMERICAN BURN ASSOCIATION 2019
A=ADULT, P=PEDIATRIC
LEVEL 1 ACS ADULT TRAUMA VERIFIED

ABA BURN CENTERS

THERE ARE 34 ABA VERIFIED BURN CENTERS THAT ARE NOT ACS ADULT LEVEL ONE NOTED ON PREVIOUS CHART. THE 34 ARE NOTED BELOW.

Name	City	State	Type	Name	City	State	Type
SHRINERS – NORTHERN CA	SACRAMENTO	CA	P	NORTH CAROLINA JAYCEE	CHAPEL HILL	NC	A & P
GROSSMAN BURN CNTR	WEST HILLS	CA	A	CHILDRENS AKRON	AKRON	OH	A & P
TORRANCE MEMORIAL MC	TORRANCE	CA	A & P	NATIONWIDE CHILDRENS	COLUMBUS	OH	P
BOTHIN BURN CNTR	SAN FRANCISCO	CA	A & P	SHRINERS – CINCINNATI	CINCINNATI	OH	P
CONNECTICUT BURN CNTR	BRIDGEPORT	CT	A & P	OREGON BURN CENTER	PORTLAND	OR	A & P
DOCTORS HOSPITAL	AUGUSTA	GA	A & P	LEHIGH VALLEY	ALLENTOWN	PA	A & P
U OF CHICAGO	CHICAGO	IL	A & P	HULNICK BURN CENTER	PHILADELPHIA	PA	P
SUMNER KOCH BURN CNTR	CHICAGO	IL	A & P	NATHAN SPEARE CNTR	CHESTER	PA	A
VIA CHRISTI REGIONAL MC	WICHITA	KS	A & P	UPMC MERCY	PITTSBURGH	PA	A
BATON ROUGE BURN CNTR	BATON ROUGE	LA	A	FIREFIGHTERS REGIONAL	MEMPHIS	TN	A
SHRINERS – BOSTON	BOSTON	MA	P	SHRINERS – GALVESTON	GALVESTON	TX	P
SUMNER REDSTONE BURN	BOSTON	MA	A	HARNAR BURN CENTER	LUBBOCK	TX	A & P
JOHNS HOPKINS BURN CNTR	BALTIMORE	MD	A & P	FORT SAM HOUSTON	SAN ANTONIO	TX	A
CHILDRENS MICHIGAN	DETROIT	MI	P	EVANS-HAYNES BURN CNTR	RICHMOND	VA	A & P
ST ELIZABETH	LINCOLN	NE	A & P	U OF WASHINGTON	SEATTLE	WA	A & P
ST BARNABAS	LIVINGSTON	NJ	A & P	MERCY HOSPITAL	ST LOUIS	MO	A & P
HEARST BURN CENTER	NEW YORK	NY	A & P	ROCHESTER MC	ROCHESTER	NY	A & P

AMERICAN BURN ASSOCIATION 2019
A=ADULT, P=PEDIATRIC

AMERICAN BURN ASSOCIATION VERIFIED BURN CENTERS IN CALIFORNIA = 10

LOS ANGELES COUNTY
    LAC + USC MEDICAL CENTER (TRAUMA CNTR)
    TORRANCE MEMORIAL MEDICAL CENTER
    GROSSMAN BURN CENTER AT WEST HILLS HOSPITAL (ADULT)
ORANGE COUNTY – UC IRVINE MC (TRAUMA CNTR)
SAN DIEGO COUNTY – UC SAN DIEGO MC (TRAUMA CNTR)

NORTHERN CALIFORNIA = 5
    SACRAMENTO – SHRINERS CHILDREN HOSPITAL (PEDIATRIC)
    SACRAMENTO – UC DAVIS MEDICAL CENTER (TRAUMA CNTR)
    SAN FRANCISCO – ST. FRANCIS MEMORIAL HOSPITAL
    SAN JOSE – SANTA CLARA VALLEY MC (TRAUMA CNTR)
    FRESNO – COMMUNITY REGIONAL

ADULT AND PEDIATRIC UNLESS NOTED
AMERICAN BURN ASSOCIATION 2019

HOSPITAL ORIENTATION

MOST PEOPLE ARE BORN IN A HOSPITAL AND DIE IN A HOSPITAL –

AND TRY TO STAY AWAY THE REST OF THE TIME

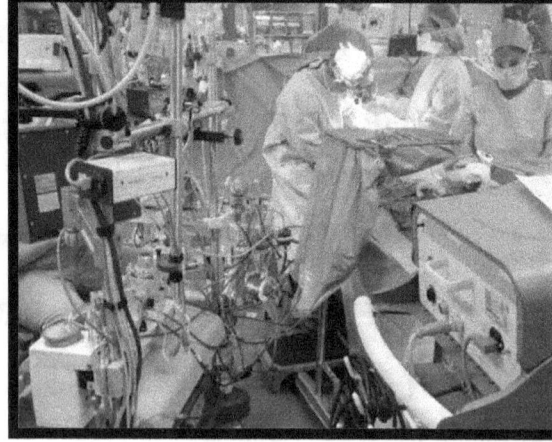

HOSPITAL ORIENTATION

ANY VISIT TO A HOSPITAL AS A VISITOR IS
FULL OF STRESS

HOSPITAL ORIENTATION

THANK YOU FOR VOLUNTEERING TO SERVE INSIDE THE HOSPITAL ENVIRONMENT

HOSPITALS = JAILS

HOSPITALS ARE AS CLOSE AS MANY OF US COME TO "JAIL" AS A PATIENT–

- TOLD WHEN TO EAT
- TOLD WHAT TO EAT
- TOLD WHEN TO GET OUT OF BED
- BATHROOM – TOLD WHEN AND HOW
- ESCORTED EVERYWHERE

HOSPITALS – UNSCHEDULED

WE DO NOT SCHEDULE TO BE ADMITTED TO AN EMERGENCY DEPARTMENT
STRESS

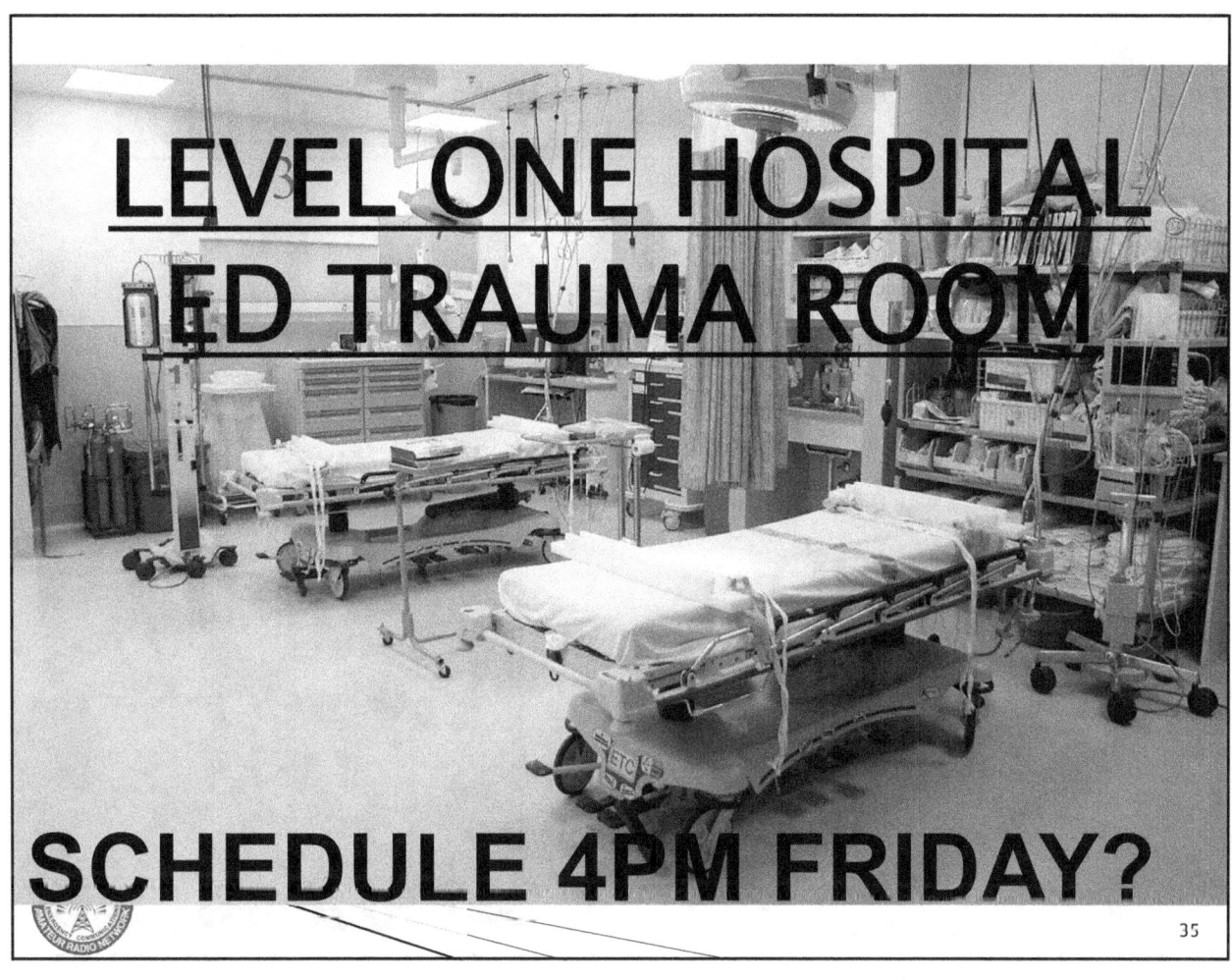

HOSPITALS – UNSCHEDULED

WE DO NOT SCHEDULE AN ACCIDENT AND EMERGENCY SURGERY

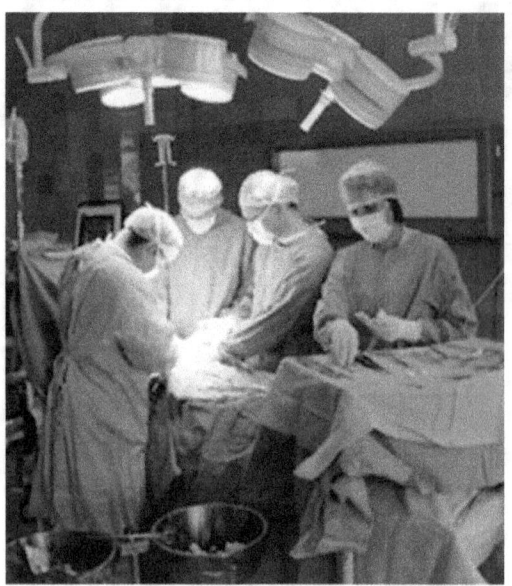

HOSPITAL REGULATORY OVERSIGHT

The Joint Commission (TJC) and the Occupational Safety and Health Administration (OSHA) regulations direct health care facilities to maintain a safe environment.

State and local regulations also require a safe environment.

HOSPITAL REGULATORY OVERSIGHT

There are numerous types of hospital regulatory reviews and surveys. These include:
- California state licensure surveys
- State and federal complaint surveys
- Accreditation surveys – TJC
- Medicare (CMMS) certification surveys
- HIPAA Field Surveys
- Post-survey revisit surveys
- Preoccupancy surveys – new hospitals
- New program surveys
- Life safety code survey – fire safety

HOSPITAL REGULATORY OVERSIGHT

FAILURE OF PERFORMANCE DETERMINED BY REGULATORY VISITS:

MAY RESULT IN
LOST REVENUES,
HOSPITAL CLOSURE
OR
CRIMINAL CHARGES.

HOSPITAL VOLUNTEER REGULATIONS

All staff (including but not limited to volunteers, per diem staff and contractors) who are working in the hospital **must adhere to policies and procedures of the hospital.**

The hospital must provide for adequate orientation, supervision and evaluation of the activities of all staff.

HOSPITAL ORIENTATION REQUIRED

- HOSPITAL SAFETY
- HOSPITAL INFECTION CONTROL
- HOSPITAL COMPLIANCE – HIPAA, ETC.
- ROLE AND RESPONSIBILITIES – RADIO OPERATOR
- SECURITY AWARENESS
- EMERGENCY PREPAREDNESS LIAISON
- PARKING, CAFETERIA AND OTHER BASICS
- HOSPITAL ACCESS
  - PHOTO ID BADGE
  - NORMAL HOURS –
  - AFTER HOURS (EMERGENCY DEPARTMENT)
- RADIO ROOM LOCATION AND ACCESS
- HCC LOCATION

VOLUNTEER RESPONSIBILITIES

- POSITIVE ATTITUDE
- TEAM PLAYER
- VERBAL COMMUNICATIONS SKILLS
- DRESS CODE
- FOLLOW HOSPITAL POLICIES
  (TOBACCO FREE, NO PARKING, NO SMOKING)
- ACTIVE PARTICIPATION
- CONFIDENTIALITY
- COMPLIANCE
- MAINTAIN RADIO OPERATOR COMPETENCY

PROFESSIONALISM

- **KPARN Radio Operators should be Professional in action & demeanor.**
- Look Professional
  - Dress Code – clean clothes – groomed
  - Name Badge
  - Appropriate Language
- Model Professional Behavior
  - What is Best for KP, KPARN and Amateur Radio
  - Work Together

BE RESPECTFUL

- Treat Everyone with Respect and Courtesy
- Keep Voice Down – Appropriate to Situation
- Communicate in a respectful and professional manner
- Positive nonverbal communication
- Be Pleasant and Welcoming
- Respect Patient and Family Privacy

BE RESPECTFUL

Remember,
Patients Do Not Feel Good,
Families are Concerned and Stressed.
Now there is a Disaster Compounding and Complicating.

HIPAA & CONFIDENTIALITY

Hospital Insurance Portability and Accountability Act (HIPAA)

What you <u>SEE</u> here and what you <u>HEAR</u> here must <u>REMAIN</u> here when you <u>LEAVE</u> here.

HIPAA & CONFIDENTIALITY

HIPAA is federal law that protects individuals' privacy and security of health information

HIPAA Privacy Rule identifies permitted use and disclosure of **Protected Health Information (PHI)**

HIPAA penalties include possible criminal and/or civil penalties

Violating HIPAA has resulted in termination

HIPAA includes on-site inspections

PHI is kept secure and confidential at all times

HOSPITAL SAFETY

SAFETY IS NUMBER ONE PRIORITY

- DRIVE SAFELY – THE DISASTER WILL LAST
- HOSPITAL PHOTO ID BADGE
- PROPER ATTIRE
  - COLLARED SHIRTS - LONG PANTS - APPROPRIATE FOOTWEAR
- PROPER ATTITUDE (MBA'S, PhD's, MD'S)
- PROPER LANGUAGE – COMMUNICATE EFFECTIVELY
- CALM – PROPER BEHAVIOR
- OBSERVE ALL SAFETY PROCEDURES
- MAINTAIN KPARN STATION – CLEAN AND ORGANIZED
- **RADIO OPERATORS SHOULD BE INTEGRATED WITH AND ORIENTATED TO HOSPITAL – NOT ASSIGNED AT TIME OF DISASTER**

HOSPITAL SAFETY

- RADIO OPERATORS DO **NOT** REPORT TO HAZARDOUS LOCATIONS
- RADIO OPERATORS ARE **NOT** INVOLVED WITH DIRECT PATIENT CARE
- HOSPITAL COMMAND CENTER (HCC) AND HOSPITAL RADIO ROOMS ARE IN "COLD ZONES"
- **IF NOT SURE ASK**
- FOLLOW DIRECTIONS OF DISASTER LEADERSHIP – ALL ELSE FAILS – FOLLOW THE CROWD – **"GRAB A TIE"**
- WORK TO MAINTAIN COMMUNICATIONS

(TELL NET CONTROL WHAT IS GOING ON)

HOSPITAL EMERGENCY CODES

WHAT DO I DO IN EVENT OF AN EMERGENCY IN THE HOSPITAL?

AS RADIO OPERATOR IN PUBLIC SPACE – FOLLOW DIRECTIONS PROVIDED TO PUBLIC

AS RADIO OPERATOR ASSIGNED TO HCC – FOLLOW DIRECTIONS PROVIDED BY INCIDENT COMMAND

Amateur Radio Hospital Orientation

HOSPITAL EMERGENCY CODES

THE CALIFORNIA HOSPITAL ASSOCIATION HAS PUBLISHED A LIST OF "STANDARDIZED" HOSPITAL EMERGENCY CODES. NOT ALL HOSPITALS FOLLOW THIS STANDARD. VALIDATE THE EMERGENCY CODES FOR YOUR HOSPITAL AS PART OF ON-SITE ORIENTATION.

Code	Description
CODE RED	Fire
CODE BLUE	Adult Medical Emergency
CODE WHITE	Pediatric Medical Emergency
CODE PINK	Infant Abduction
CODE PURPLE	Child Abduction
CODE YELLOW	Bomb Threat
CODE GRAY	Combative Person
CODE SILVER	Person with a Weapon and/or Active Shooter and/or Hostage Situation
CODE ORANGE	Hazardous Material Spill / Release
CODE GREEN	Missing High Risk Patient
CODE TRIAGE – EMERGENCY ALERT	Limited activation of selected key personnel for potential incident
CODE TRIAGE – INTERNAL	Activate Emergency Operations Plan for Internal Incident
CODE TRIAGE – EXTERNAL	Activate Emergency Operations Plan for External Incident

HOSPITAL FIRE SAFETY

All hospitals have fire response procedures that all staff and volunteers must know and be prepared to implement in order to protect patients, co-workers, visitors, themselves and property from real or suspected fires.

- Fire doors, corridors and stairs must always remain clear, unobstructed and free from storage to allow for safe evacuation during an emergency.
- There are always two different exit routes out of an area or floor.
- Evacuation routes, corridors and stairwells are clearly marked by "EXIT" signs.
- Do not use elevators during a fire. Use the stairs.
- In patient care areas, it is preferable to "defend-in-place" by closing doors. If evacuation is necessary, evacuate horizontally, staying on the same floor, but proceeding past a set of fire doors in the corridor.

HOSPITAL FIRE SAFETY

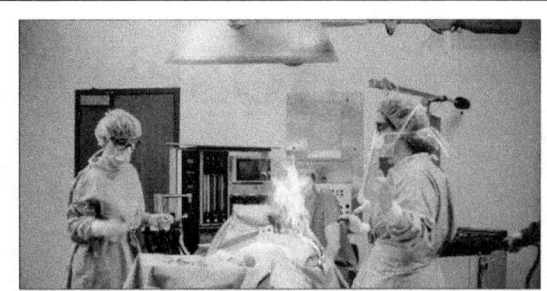

- Know the location of **Fire Safety Equipment** near your "work area" – typically the Radio Room.
- Know the location of **Fire Extinguishers** near the Radio Room and HCC
- Know the location of two exterior **Exits** near the Radio Room and HCC
- Know the location of the **Manual Fire Alarm Pull Station** near the Radio Room and HCC
- Include in On-Site Orientation the **Phone Number** to call on campus to report a Fire

Amateur Radio Hospital Orientation

HOSPITAL FIRE SAFETY
Reporting A Fire: R.A.C.E.

R – Rescue – remove patients and people from immediate danger

A – Alarm – pull the fire alarm station, call out "fire!" and dial the emergency number extension (hospital specific)

C – Contain – the fire and smoke by closing doors

E – Extinguish – small fire with fire extinguisher only if safe to do so. If fire is not small EVACUATE area.

HOSPITAL FIRE SAFETY

Reporting A Fire: P.C.A.E.

P – Patients are to be removed from immediate danger

C – Contain the fire by closing doors

A – Activate the alarm system by pulling the fire alarm box and by dialing extension (hospital specific)

E – Extinguish the fire with fire extinguisher only if safe to do so

P.C.A.E. is used in City of Los Angeles

HOSPITAL EMERGENCY POWER

- RED RECEPTACLE IS ON EMERGENCY GENERATOR
- WHEN NORMAL POWER FAILS BETWEEN TWO OR MORE (SITE DEPENDENT) GENERATORS START AND SYNCHRONIZE POWER OUTPUT.
  - THESE ARE LARGE – SIZE OF TRAIN ENGINE - GENERATORS IN NEW KP FACILITIES
  - PER CODE 96 HOURS (+1.33 NFPA RESERVE) OF FUEL IS REQUIRED FOR NEW CONSTRUCTION
  - TYPICALLY GENERATORS RUN AT LESS THAN 50% LOAD
  - SYNCHRONIZATION ALLOWS ONE GENERATOR TO FAIL WITH MINIMAL EFFECT ON HOSPITAL LOAD.

HOSPITAL EMERGENCY POWER

- AUTOMATIC TRANSFER SWITCHES (ATS) CHANGE LOAD FROM NORMAL TO EMERGENCY POWER
  - THERE MAY BE DOZENS OF ATS IN HOSPITAL
  - THESE ARE THE SIZE OF LARGE REFRIGERATORS
  - PER CODE TRANSFER IS LESS THAN 15 SECONDS
- RED LIGHT SWITCH HANDLE OR PLATE SIGNIFIES GENERATOR POWERED LIGHT
- SOME LIGHTS HAVE BUILT IN BATTERY BACK-UP TO PREVENT TOTAL BLACK DURING 15 SECOND TRANSFER (OR)

HOSPITAL RESPONSIBILITIES
DISASTER PREPAREDNESS

- **PROTECT PATIENTS**
- **PROTECT STAFF**
- **PROTECT FACILITY**

- Reinforced by Regulations and the Potential to Withhold Funds

DISASTER PREPAREDNESS HOSPITALS ARE UNIQUE

- Typically, hospitals are immediately involved in disasters.

DISASTER PREPAREDNESS HOSPITALS ARE UNIQUE

- Typically, hospitals are immediately involved in disasters.
- **Hospitals are open 24 by 7 and involved long before Emergency Management, Health Departments and other "support" agencies.**
  - The Joplin Hospital evacuation took about an hour, well before other organizations were able to assist.

DISASTER PREPAREDNESS HOSPITALS ARE UNIQUE

- Typically, hospitals are immediately involved in disasters.
- Hospitals are open 24 by 7 and involved long before Emergency Management, Health Departments and other "support" agencies.
- **Hospitals are a "Beacon" of light, warmth, food and comfort to the community – SAFE HAVEN (overwhelming numbers)**

DISASTER PREPAREDNESS HOSPITALS ARE UNIQUE

- Typically, hospitals are immediately involved in disasters.
- Hospitals are open 24 by 7 and involved long before Emergency Management, Health Departments and other "support" agencies.
- Hospitals are a "Beacon" of light, warmth, food and comfort to the community – **SAFE HAVEN**
- **Hospital license and reimbursement linked to disaster planning.**
  SIGNIFICANT REGULATORY COMPONENT

DISASTER PREPAREDNESS
HOSPITALS ARE UNIQUE

- Typically, hospitals are immediately involved in disasters.
- Hospitals are open 24 by 7 and involved long before Emergency Management, Health Departments and other "support" agencies.
- Hospitals are a "Beacon" of light, warmth, food and comfort to the community – **SAFE HAVEN**
- Hospital license and reimbursement linked to disaster planning. **REGULATORY**
- **PREPLANNING AND ORIENTATION – DON'T JUST SHOW UP**
- *Radio operators should be integrated with and orientated to hospital – not assigned at disaster.*

Disaster and Mass Casualty Program

CA TITLE 22

(a) A written disaster and mass casualty program shall be developed and maintained in consultation with representatives of the medical staff, nursing staff, administration and fire and safety experts. The program shall be in conformity with the California Emergency Plan of October 10, 1972 developed by the State Office of Emergency Services and the California Emergency Medical Mutual Aid Plan of March 1974 developed by the Office of Emergency Services, Department of Health. The program shall be approved by the medical staff and administration. A copy of the program shall be available on the premises for review by the Department.

(b) The program shall cover disasters occurring in the community and widespread disasters. It shall provide for at least the following:

(1) Availability of adequate basic utilities and supplies, including gas, water, food and essential medical and supportive materials.

(2) An efficient system of notifying and assigning personnel.

(3) Unified medical command.

(4) Conversion of all usable space into clearly defined areas for efficient triage, for patient observation and for immediate care.

(5) Prompt transfer of casualties, when necessary and after preliminary medical or surgical services have been rendered, to the facility most appropriate for administering definite care.

(6) A special disaster medical record, such as an appropriately designed tag, that accompanies the casualty as he is moved.

(7) Procedures for the prompt discharge or transfer of patients already in the hospital at the time of the disaster who can be moved without jeopardy.

(8) Maintaining security in order to keep relatives and curious persons out of the triage area.

(9) Establishment of a public information center and assignment of public relations liaison duties to a qualified individual. Advance arrangements with communications media will be made to provide organized dissemination of information.

(c) The program shall be brought up-to-date, at least annually, and all personnel shall be instructed in its requirements. There shall be evidence in the personnel files, e.g., orientation checklist or elsewhere, indicating that all new employees have been oriented to the program and procedures within a reasonable time after commencement of their employment.

(d) The disaster plan shall be rehearsed at least twice a year. There shall be a written report and evaluation of all drills.

HOSPITAL COMMAND CENTER (HCC)

HOSPITAL COMMAND CENTER (HCC)

AMATEUR RADIO OPERATORS ARE TYPICALLY THE ONLY NON-ADMINISTRATIVE STAFF IN THE HOSPITAL COMMAND CENTER. PROFESSIONALISM AND CONFIDENTIALITY IS KEY TO RADIO OPERATOR ROLE

SEE SOMETHING? SAY SOMETHING!

- Program is designed to encourage all staff to report suspicious circumstances to Hospital Security for prompt response.

If you witness an unusual incident or occurrence, report it to Hospital Security immediately.

SEE SOMETHING? SAY SOMETHING!

How Does It Involve Me?

- You know what is normal, and more importantly, what is not.
- Be aware of suspicious objects, packages or vehicles.
- If you hear a threat by someone that could be credible, take it seriously and report it.
- Take note of your surroundings
- Look out for unusual conduct or behavior

HOSPITAL SAFETY

While all patient care areas are considered limited access – specific hospital areas are "secure" from access.

No Radio Operator should need to be in these "secure" areas without appropriate escort.

Typical Secure areas include: MRI Suite, Nuclear Medicine, Operating Rooms, Medication Rooms, Server Farm, Intensive Care Unit (ICU), and Pharmacy

HOSPITAL SAFETY

SPEAK UP!!

IF SOMETHING DOES NOT SEEM SAFE TO YOU, INFORM AN APPROPRIATE HOSPITAL STAFF MEMBER

IF IT IS NOT SAFE – DON'T DO IT

REPORT ANY ACCIDENT IN WHICH YOU ARE INVOLVED

HOSPITAL ORIENTATION

Some amateur radio groups expect to be in clinical patient care areas. KPARN radio operators should not be in a patient care area unless the operator has been completely orientated to the patient care environment.

KPARN radio operators are NOT orientated to be in a patient care area.

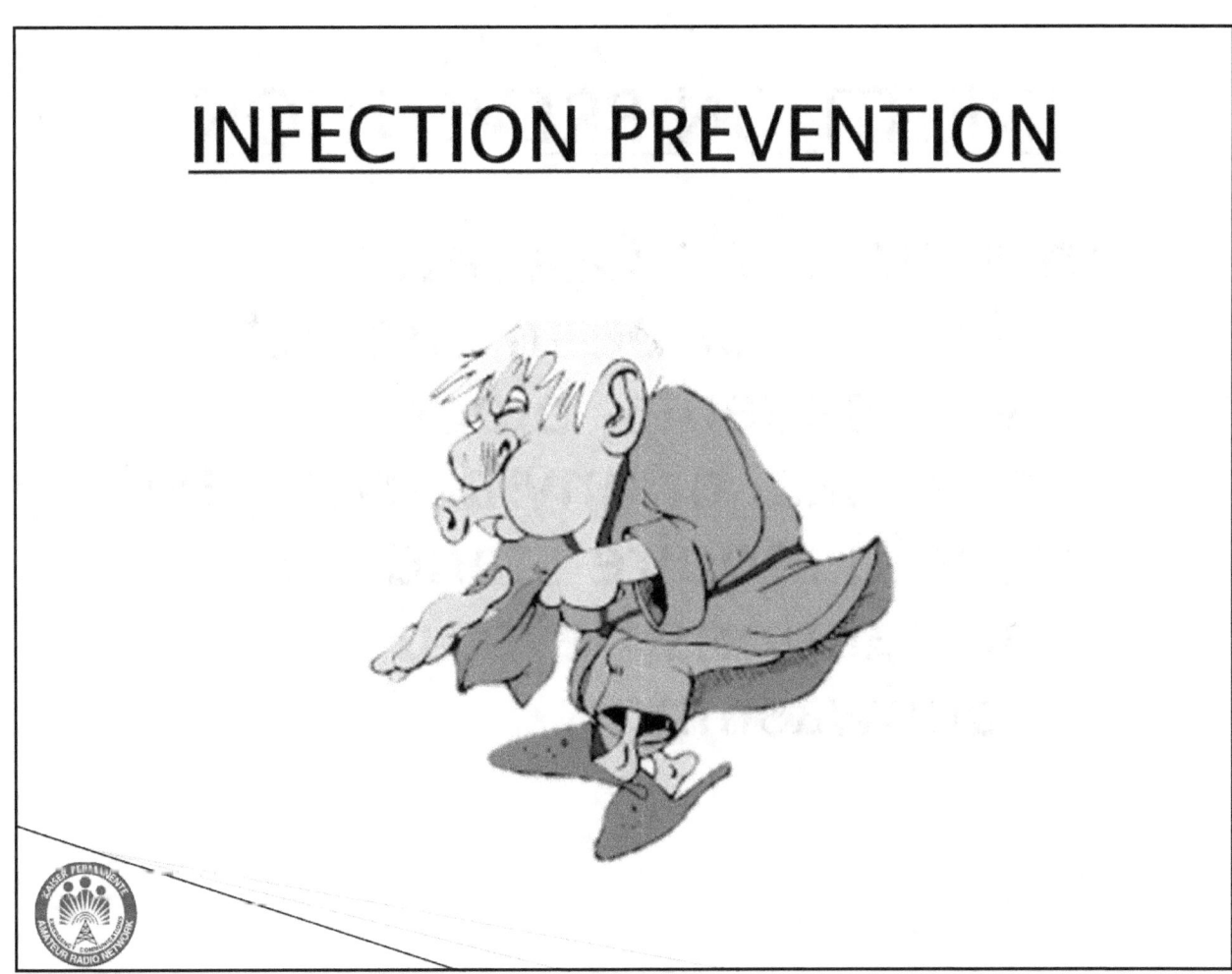

INFECTION PREVENTION

Amateur Radio Operators should not be in contact with patients, blood, syringes, or any activity that would require standard precautions beyond handwashing.

INFECTION PREVENTION

Implementation and **adherence** to infection prevention practices are the keys to preventing the transmission of infectious disease to yourself, family members, patients, coworkers and community members in general.

Hand hygiene is the **single** most effective means of preventing the spread of all infections among hospital patients, staff and visitors.

CDC

INFECTION PREVENTION HAND HYGIENE

The proper method to wash your hands using soap and water:
- Wet your hands
- Apply soap and scrub for fifteen seconds
- Wash all areas of the hand including between the fingers
- Rinse well
- Dry hands with paper towel
- Use paper towel to turn of faucet
- Use paper towel to turn door knob

INFECTION CONTROL

- RADIO OPERATORS SHOULD BE CURRENT WITH VACCINES, FLU SHOTS, TETNAUS, HEP B, TB TEST, ETC ROUTINELY PROVIDED TO ADULTS
- ***WASH YOUR HANDS***
- USE HAND SANITIZERS (THEY ARE ALL OVER HOSPITAL)
- DO NOT USE PERSONAL PROTECTIVE EQUIPMENT (PPE) (FACE MASKS, PAPR'S, ETC) UNLESS CURRENTLY CERTIFIED, FIT TESTED – AND MEDICALLY CLEARED

INFECTION CONTROL

ANNUAL FLU SHOT

IF SICK – STAY HOME

DO NOT CONTAMINATE THE HCC

**THANK YOU
FOR SUPPORTING
AMATEUR RADIO
AND
HOSPITAL EMERGENCY
PREPAREDNESS**

KAISER PERMANENTE

Kaiser Permanente (KP) is recognized as one of America's leading health care providers and not-for-profit health plans. Kaiser Permanente currently serves more than 10 million members in eight states and the District of Columbia.

10 million is more than population in 42 states

KAISER PERMANENTE MISSION

Kaiser Permanente (KP) exists to provide high-quality, affordable health care services and to improve the health of our members and the communities we serve.

KAISER PERMANENTE VISION

Kaiser Permanente vision is to consistently provide high quality affordable healthcare in an easy and convenient manner with a personal touch.

KP HOSPITALS SOUTHERN CALIFORNIA

NAME	CITY	COUNTY	BEDS
LOS ANGELES	LOS ANGELES	LOS ANGELES	460
FONTANA	FONTANA	SAN BERNARDINO	450
SAN DIEGO - ZION	SAN DIEGO	SAN DIEGO	406
DOWNEY	DOWNEY	LOS ANGELES	352
WOODLAND HILLS	WOODLAND HILLS	LOS ANGELES	280
BALDWIN PARK	BALDWIN PARK	LOS ANGELES	272
WEST LA	LOS ANGELES	LOS ANGELES	265
ANAHEIM	ANAHEIM	ORANGE	262
SOUTH BAY	HARBOR CITY	LOS ANGELES	257
SAN DIEGO MC	SAN DIEGO	SAN DIEGO	253
RIVERSIDE	RIVERSIDE	RIVERSIDE	226
PANORAMA CITY	PANORAMA CITY	LOS ANGELES	218
IRVINE	IRVINE	ORANGE	207
ONTARIO	ONTARIO	SAN BERNARDINO	176
MORENO VALLEY	MORENO VALLEY	RIVERSIDE	94
MENTAL HEALTH CENTER	LOS ANGELES	LOS ANGELES	68

OSHPD 2019

"COMMUNITY" OR LEVEL 2 NICU KP SOUTHERN CALIFORNIA

KP Anaheim MC
KP Baldwin Park MC
KP Downey MC
KP Fontana MC
KP Panorama City MC
KP Riverside MC
KP San Diego MC
KP West LA MC
KP Woodland Hills MC

CA DHCS

ED VISITS
KP SOUTHERN CALIFORNIA

KP FONTANA	127,000
KP ANAHEIM	105,000
KP ZION	100,000
KP DOWNEY	96,000
KP BALDWIN PARK	79,000
KP WEST LA	74,000
KP LAMC	65,000
KP PAN CITY	65,000
KP SOUTH BAY	64,000
KP RIVERSIDE	47,000
KP MORENO VALLEY	43,000
KP WOODLAND HILLS	40,000
KP ONTARIO	
KP IRVINE	
KP SDMC	

CA OSHPD – 2017

Duane Mariotti, WB9RER

Duane is an Electrical Engineer specializing in Biomedical integration. He has been involved with emergency communications for almost thirty years as an engineer, responder and policy leader. He has worked as an EMT, Paramedic, Firefighter and been a college level educator for biomedical technology and hazardous materials. Following 9/11 he served on several state and national committees improving hospital emergency preparedness and communications systems. He has published numerous articles and book chapters related to hospital technology and emergency communications. Most recently Duane was a Clinical Engineer for Kaiser Permanente in Southern California integrating medical technology into new facilities and patient safety initiatives. He is currently the Volunteer Coordinator of the Kaiser Permanente Amateur Radio Network (KPARN).

QUIZ

Self-Test Your Hospital Knowledge

Question 1

Which of the following could happen to someone who violates HIPAA regulations and shares Protected Health Information (PHI) for reasons not associated with their volunteer Radio Operator duties?

A.) Dismissal from Volunteering
B.) Civil Lawsuit
C.) Criminal Prosecution
D.) All of the above

Question 2

One way we all help to lower the risk of spreading infection is to properly wash our hands.

A.) True
B.) False

Question 3

Which of the following are examples of Protected Health Information (PHI) and would be covered by HIPAA regulations?

A.) A photo of a patient
B.) The written address of a patient
C.) A patient's email address
D.) Speaking the name of the patient on an elevator
E.) All of the above

Question 4

Employees, Volunteer Radio Operators and other staff must keep patient and hospital information confidential.

A.) True
B.) False

Question 5

Fire safety rules include all the following except:

A.) Use elevators in the event of fire.
B.) Keep hallways clear
C.) Do not block exits, fire alarms or prop doors open
D.) Do not store supplies or boxes on the floor

Question 6

Sensitive areas are areas of restricted access open only to authorized personnel.

A.) True
B.) False

Amateur Radio Hospital Orientation

Question 7

Which are dress code requirements?

A.) Employees/volunteers are required to wear identification badges at all times while on duty
B.) Employees/ volunteers are expected to be professional in appearance
C.) Attire shall be modest, safe, and clean while on duty
D.) All of the above

Question 8

HIPAA is Federal Law aimed at protecting confidentiality and security of health data.

A.) True
B.) False

Question 9

Which of the following is NOT one of the four diseases most likely to occur as a result of bioterrorism?

A.) Anthrax
B.) Chickenpox
C.) Plague
D.) Botulism

Question 10

Volunteer Radio Operators must wear the provided KP identification issued badge at all times while officially "working" on campus.

A.) True
B.) False

Question 11

The regulatory agencies which hospitals must be compliant with include:

A.) TJC –The Joint Commission on the Accreditation of Healthcare Organization
B.) OSHA – Occupational Safety & Health Administration
C.) OSHPD – California Office of Statewide Health Planning and Development
D.) All of the above

Amateur Radio Hospital Orientation

Question 12

Identification badges must be worn and visible at all times.

A.) True
B.) False

Question 13

In the event of a power failure, all critical and life saving equipment must be plugged in _____.

A.) Safely, out of reach
B.) To emergency battery powered backpacks
C.) To red, emergency power receptacles
D.) To white, normal power receptacles

This includes KPARN radios.

Question 14

Hospitals are committed to providing employees and Radio Operator volunteers a workplace that is free from acts or threats of violence.

A.) True
B.) False

Question 15

Which of the following are considered sensitive areas:

A.) Medication rooms
B.) Emergency room and Critical Care areas
C.) Pharmacy and Medical Records
D.) MRI Suite
E.) All of the above

Question 16

HIPAA violation can result in civil and / or criminal penalties.

A.) True
B.) False

Question 17

Examples of protected health information (PHI) are:

A.) Patient names
B.) Medical Record numbers
C.) Social Security numbers
D.) All of the above

Question 18

Amateur radio operators assigned to hospitals are specifically orientated to work with patients in the clinical environment.

A.) True
B.) False

Question 19

If a fire occurs, we follow the acronym R–A–C–E. Which of the following are correct?

A.) Rescue – Alarm – Contain – Extinguish or Evacuate
B.) Run – Alarm – Call – Evaluate
C.) Rescue – Assist – Call – Evacuate

Question 20

By regulation, everyone who provides service in a hospital paid, volunteer, part time, or contractor must be orientated to the hospital environment.

A.) True
B.) False

Question 21

Students, volunteers and contract workers should do which of the following?

A.) NOT smoke on hospital property (outside of designated areas) Including e-cigarettes
B.) NOT bring a weapon in the hospital as they are prohibited.
C.) Wear their ID badge while at the hospital
D.) All of the above

Amateur Radio Hospital Orientation

Question 22

The largest hospital (by number of inpatient beds) in California has over one thousand beds.

A.) True
B.) False

Question 23

In the memory device R-A-C-E, R stands for:

A.) Remember to close doors and windows
B.) Remember to pull the fire alarm
C.) Rescue people in immediate danger
D.) Run for the door and leave the building

Question 24

The total number of acute care beds available in the United States has doubled today compared with 1975.

A.) True
B.) False

Question 25

The most common mode of transmission of pathogens is:

A.) Waste containers
B.) Blood cultures
C.) X-ray equipment
D.) Hands

Question 26

There are more than two dozen Level One Trauma Centers in Southern California.

A.) True
B.) False

Question 27

The largest hospital (by number of inpatient beds) in the United States is located in:

A.) California
B.) Florida
C.) Texas
D.) Pennsylvania

Question 28

All Level One Trauma Centers in Southern California are verified American Burn Association Burn Centers.

A.) True
B.) False

Question 29

The most comprehensive level of care for an accredited trauma center is level:

A.) Four
B.) Three
C.) Two
D.) One

Question 30

Radio operators should be integrated with and orientated to hospital – not assigned at disaster.

A.) True
B.) False

Amateur Radio Hospital Orientation

Question 31

An on-site regulatory review may be conducted by:

A.) TJC
B.) CMMS
C.) Fire Marshal
D.) All the above

Question 32

All new hospital staff including physicians, contractors, volunteers and employees typically attend hospital orientation.

A.) True
B.) False

Question 33

The most comprehensive level of care for an accredited neonatal center is level:

A.) Four
B.) Three
C.) Two
D.) One

Question 34

A Level 1 Trauma Center has 24-hour in-house coverage by general surgeons

A.) True
B.) False

Question 35

The State with the most licensed acute care hospitals is:

A.) California
B.) New York
C.) Texas
D.) Pennsylvania

Question 36

Personal Protective Equipment (PPE) (N95 Face Masks, PAPR'S, etc.) should not be used unless wearer is currently CERTIFIED, FIT TESTED – and MEDICALLY CLEARED.

A.) True
B.) False

Question 37

Personal Protective Equipment (PPE) includes:

A.) Face Shields
B.) PAPR's
C.) Latex Free Gloves
D.) N-95 Mask
E.) All the Above

Question 38

Like 9-1-1 there is a common standardized internal emergency activation phone number for all hospitals. There are common emergency codes for fire, disasters, etc. shared and consistent among hospitals.

A.) True
B.) False

Amateur Radio Hospital Orientation

Question 39

For your emergency communications team response area, the closest ABA Verified Adult Burn Center is

_____.

Question 40

Hospital orientation is typically configured to assist with interactions and activities related to patient care.

A.) True
B.) False

Question 41

For your emergency communications team response area, the closest adult ACS Level One Trauma Center is:

_____.

Question 42

If a hospital fails The Joint Commission (TJC) survey, the hospital may jeopardize its state license and / or have financial income curtailed through insurance carriers.

A.) True
B.) False

Question 43

Hospitals may have numerous on-site regulatory inspections. Failure to meet inspection standards may result in:

A.) Lost Revenues
B.) Hospital Closure
C.) Criminal Charges
D.) All the above

Question 44

You must **never** touch a switch with red cover plate or handle

A.) True
B.) False

Question 45

For your emergency communications team response area, the closest state or ACS verified Adult Trauma Center any level is:

_____.

Question 46

In the "standardized" hospital emergency code Code Gray means "missing high risk elderly patient".

A.) True
B.) False

Question 47

For your emergency communications team response area, the closest verified pediatric ACS Level One Trauma Center is:

_____.

Question 48

The busiest emergency department in the USA is in California.

A.) True
B.) False

Question 49

The busiest emergency department in the USA has over 200,000 patient contacts per year. This is equivalent to an average of one patient contact every two minutes or 600 patients per day.

A.) True
B.) False

Question 50

Hospitals are a "Beacon" of light, warmth, food and comfort to a community affected by a disaster. The local hospital is a **SAFE HAVEN.**

A.) True
B.) False

www.ingramcontent.com/pod-product-compliance
Lightning Source LLC
Chambersburg PA
CBHW080919170526
45158CB00008B/2162